Cowgirl Up! Let's Go

Cowgirls from left, are, Sheridan Chaney, Lindsi Loos, Libbi Loos, Rianna Chaney and Landri Loos

By Twins Rianna and Sheridan Chaney

Seven-year old sisters, Rianna and Sheridan Chaney, had the unique opportunity to spend a week with Libbi (13), Lindsi (8), and Landri (6) Loos on their family's ranch in Litchfield, Nebraska. The whole idea of a visit started when the Chaney twins were given an assignment n first grade by their teacher, Mrs. Lipchock. She read a book to the class entitled "Flat Stanley." In Stanley Lambchop's story a bulletin oard falls, flattens him and he's able to travel to many places experiencing great adventures with his brother, Arthur. Mrs. Lipchock had each of her students send a Flat Stanley cutout and a letter to a friend around the world. The Chaneys sent their Flat Stanley to the Loos isters in Nebraska. The Flat Stanley project was established in 1995 by Dale Hubert, third-grade school teacher in London, Ontario, Canada. It is meant to facilitate school children letter writing as they document where Flat Stanley has traveled. It was Flat Stanley's exciting ranch adventure that encouraged Rianna and Sheridan to want to experience ranch life firsthand.

To all the Farmers and Ranchers in America,

Our Cowgirl hats are off to all the farmers and ranchers who produce safe, wholesome and nutritious food for consumers. We are proud to be fifth generation farm girls, helping produce food for America.

Published by Down Under Publications

First Edition, June, 2012

ISBN 978-0-9818468-3-5

Chaneyswalkabout@aol.com
website: www.rebeccalongchaney.com
Blog: chaneytwinsagbook.blogspot.com
Edited by Rebecca Long Chaney
Photographs by Kelly Hahn Johnson
Layout and Design by Kathy Moser Stowers

Reference Material: Flat Stanley by Jeff Brown; Pictures by Scott Nash

Hey, Sheridan, I think I'm sending my Flat Stanley letter to the Loos family in Nebraska. I've always dreamed of living on a ranch. Now, our friends, Libbi, Lindsi and Landri can take photos with Flat Stanley. I hope Flat Stanley has fun at the ranch.

Like many states in America's Heartland, that's out West to you and me, there may not be a neighbor for miles, so it's fun to get mail. The youngest Loos sisters, Landri and Lindsi, are thrilled to get another pen pal letter from us.

When we first arrive we are friends right away. We have so much fun together. They have a mule named Toby and he is soooo slow.

The first morning Landri and her cousin Abbie take us right to the swine barn. Swine is a fancy name for pig. Pigs are one of our favorite animals, even if pig poop really smells! We learn the white pigs are Yorkshires and the black and white ones are Hampshires. The notches in the piglets' ears tell us their litter number and when the piglets were born.

Landri and I pick up two piglets to see if they like each other. Our best friend Brady from Maryland is with us on our ranching trip. This is the first time for him to feed a big sow. After the sow finishes her lunch it is time for her litter of piglets to have their snack.

Time to wash pigs... now, that is a challenge! Lindsi is in 4-H and will show her pigs at the county and state fair. She also shows at the World Pork Expo. See the marshmallow on the end of Landri's stick? Pigs love them and it helps teach the pigs to walk calmly so the judge gets a good view. When we get pigs next year, we are going to use marshmallows, too! Lindsi teaches us to use a show stick.

After a long day of working and "playing" on the ranch, it's boots off and a very needed break. Abbie enjoys showing us her Nintendo DS. Everything at the ranch is a new experience, even playing a video game.

We are busy every minute at the ranch, whether it is helping with all the animals, playing on the trampoline, riding bikes, or relaxing on the tree house porch after lunch. Boy, we sure are tired tonight as we all sleep in the tree house under the stars of the big western sky.

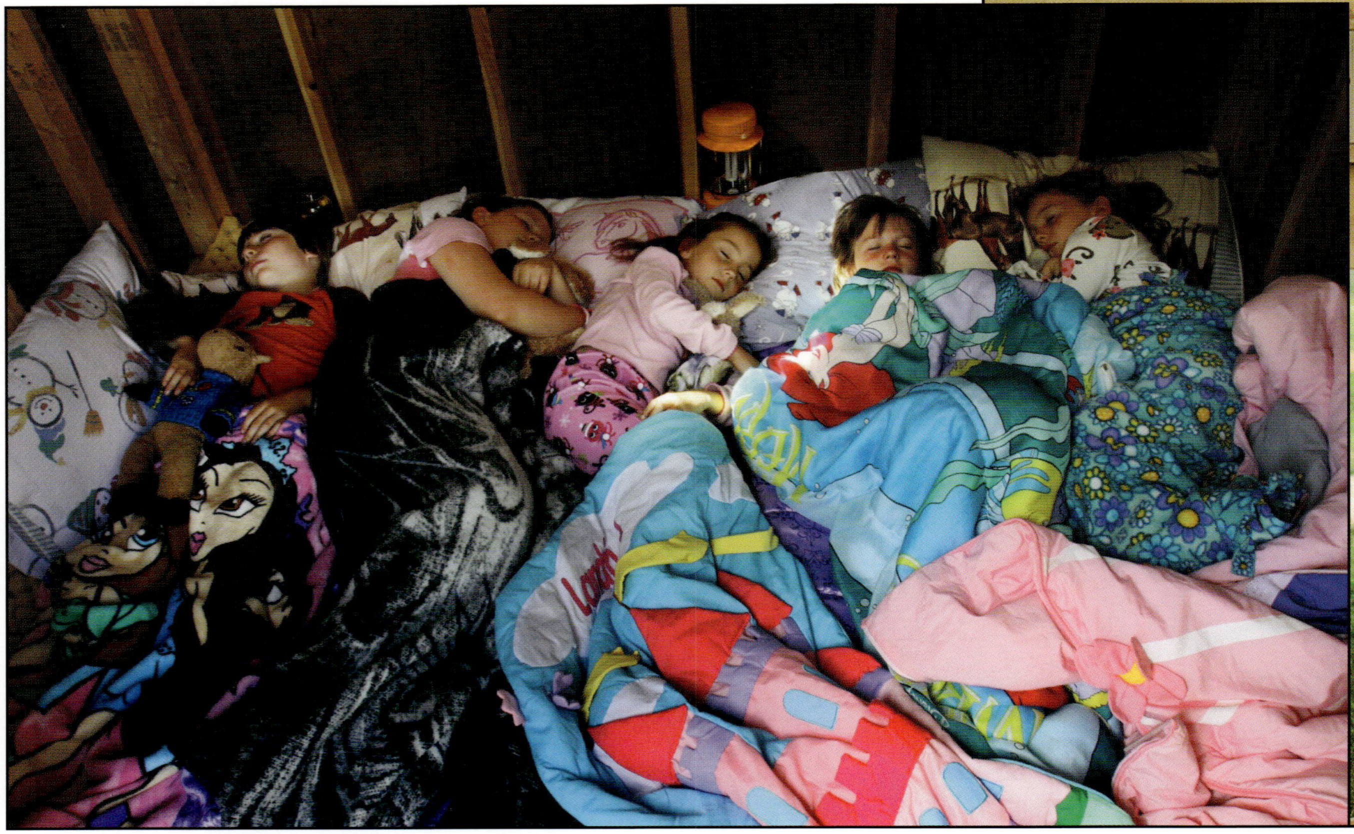

Libbi helps her dad move the cattle to the watering tank so they can all be observed. On the ranch regular checks are very important to make sure the cattle are healthy. Some ranches out West are so big the only way to check the cattle is on horseback or with airplanes — that's awesome!

Libbi, Lindsi and their dad give their horses a needed drink once the cattle are safely moved to the watering tank. The family dog, Cowgirl, is hot and thirsty, and takes a dip in the watering tank. Dogs are so funny!

"Uncle Trent" doesn't think that horseplay is a good idea, but we sure have fun at the watering tank, splashing each other on a hot, hot day. But what's cool is learning that raising beef cattle is the largest segment of American agriculture in the United States.

It's not a ranch without a herd of horses. Ranches depend on their working horses. We hope one day Mommy and Daddy will get us a horse to help round up our cattle.

Libbi goes out to gather up the horses. She is such a great rider. We love to watch her ride and bring the horses into the nearby paddock. She even gave each of us a ride on her horse, Diamond.

Brady and I might not be great goat wranglers, but we try our best. When we want the goats to go left, they run right and when we want them to go right, they go left.

This is just like piercing ears but I'm tagging this little goat's ear. A young goat is called a kid. Libbi holds the kid while I use the tagger and Rianna and Landri help calm the kid. Ear tags identify every goat because the Loos family has 200 — wow, they have lots of kids!

With grain in a bucket, Lindsi coaxes the goats toward the trailer to load them for market. The family sends their meat goats to special markets far from their ranch.

When Lindsi runs out of grain, I have more grain in a scoop so the goats follow me right up in the trailer. That grain sure makes them easy to load. It is another responsibility we learn on the ranch!

One of our most enjoyable chores on the ranch is helping Libbi wash her show steer. We fill the wash bucket with soap and water, it smells kinda funny. And knowing us, we will have more soap on us than the steer. Most kids don't know that some farm and ranch animals get baths, too.

We also get to use a livestock blower. That thing is pretty neat, especially using it on each other. The blower helps you style the steer's hair for the show ring. It's not easy grooming a steer.

Rianna, Lindsi and I ride in the back of the horse wagon. We have a blast and love Uncle Trent's big Percheron draft horse, Jim. Draft horses, like Jim, have lots of power and pull wagons in parades.

Brrr... it is a chilly evening but we are warmed by the campfire and friends we made at the ranch. We had so much fun learning about ranch life and hope we can come visit the Loos family again one day.

Hey guys, let's not miss this incredible sunset. Wow! I think this is
we are blessed to help produce safe and nutritious food for Americ

eautiful way to finish our visit. Whether we are ranchers or farmers,
ell, buckaroos, this is the end of our adventure on the ranch.

Glossary

4-H - The largest youth development organization in North America focusing on leadership, citizenship and life skills.
America's Heartland - A word to describe states that are located in the center of the United States.
Flat Stanley - A book character who takes great adventures as a writing assignment in elementary schools.
Hampshire - A special breed of pigs that originated in the United Kingdom.
Kid - A young goat.
Litter - A word describing a newborn group of piglets.
Livestock Blower - A special blower used to blow and style the hair of livestock.
Meat Goat - A goat raised specifically for the production of meat consumption.
Paddock - A small fenced-in pasture near a stable or barn.
Percheron - A special breed of draft horse known for working and pulling wagons.
Piglet - A young pig.
Ranch - Land used for raising livestock under range or grazing conditions.
Rancher - A person who owns or works on a ranch.
Show Stick - A special tool used in the show ring to guide pigs during a competition.
Sow - A mother pig.
Steer - A bull calf that has been castrated.
Swine - The scientific name for pigs.
West - All the states that are located west of the Mississippi River.
World Pork Expo - A really big competition held annually in Iowa for swine and swine enthusiasts.
Wrangler - A person who handles livestock.
Yorkshire - A special breed of pigs that originated in England.

Fun Livestock Facts

More than 97% of U.S. beef cattle farms and ranches are family owned • Beef is an excellent source of 10 essential nutrients • Beef fat and proteins help make lipstick, marshmallows, and toothpaste • Pigs provide enough artificial insulin for one diabetic for a year, plus more than 40 other medications • 800,000 ranchers and farmers raise 1 million cattle in the U.S. • There are more goats in the world than any other livestock animal • Pig heart valves are used for many human heart transplants • The hide of one beef animal can produce 11 basketballs or 32 baseballs • Goats are raised for meat, milk, and a popular cheese called feta • The first hamburger was invented in 1904 at the St. Louis World's Fair by Fletcher Davis • The most pork is produced in Iowa • The top six beef producing states are Texas, Missouri, Oklahoma, Nebraska, South Dakota and Montana

For Teachers and Parents

www.pabeef.org - The Pennsylvania Beef Council (PBC) is a non-profit organization working on behalf of 25,000 beef, dairy and veal producers in Pennsylvania. The Council seeks to promote the beef industry in Pennsylvania and educate consumers through a variety of programs and resources including nutrition research, nutrition education, Student education, beef safety research, beef quality assurance programs, meat marketing management, retail, food service, consumer promotions and dairy beef quality assurance programs. To learn more about beef quality assurance, visit www.pa-bqa.org. Lesson Plans are available free in downloadable form from PBC website.

www.ancw.org – The mission of the American National Cattlewomen is to promote and support the industry, and encourage women involved in beef and related agribusiness. They work to sustain the integrity of the beef industry through consumer awareness, education, and promotion; continuing to respect the past, living in the present and looking toward the future.

ExploreBeef.org – Explore Beef provides information on the family farmers and ranchers who raise beef, animal care, environment, safety and nutrition.

BeefitsWhatsforDinner.com – This website is the number one consumer site for beef recipes and information on preparing and serving beef.

SchoolWellness.org – This website provides a variety of tools to help teachers, school wellness leaders, health professionals and families address the nutrition and health challenges facing the youth of today.

www.beefnutrition.org – Beef nutrition is brought to you by America's Beef Producers through the Cattlemen's Beef Board and State Beef Councils. Events and seminars, media reports, materials and downloads, nutrition information, nutrition research, Dietary Guidelines Alliance, Nutrition Rich Foods Coalition, and recipes are available on this website.

www.facesofag.com – Faces of Agriculture is a nationwide, grass-roots organization designed to promote all of production agriculture by providing producers with empowering information so they can confidently champion their industry.

www.Teachfree.com – Teach free provides preschool through 12th grade educators with high-quality educational materials that supplement the curriculum. The National Cattlemen's Beef Association works collaboratively with its partners at the state level. There are state beef councils in 45 states, and they may offer free materials and assistance to teachers in their states. Most of the educational materials on teachfree have been developed as a result of this partnership.

www.4-h.org – 4-H has grown into a community of 6 million young people across America learning leadership, citizenship and life skills. 4-H can be found in every county in every state, as well as the District of Columbia; all U.S. territories and more than 98 countries around the world. 4-H'ers participate in fun, hands-on learning activities, supported by the latest research of land-grant universities.

www.FFA.org – FFA is a dynamic youth organization that is a part of agricultural education programs at middle and high schools with nearly 1/2 million enrolled. Today, student members are engaged in a wide range of curriculum and FFA activities, leading to more than 300 career opportunities in agriculture.

www.animalagalliance.org – It's never too early to start teaching your children about agriculture! The Alliance has compiled a collection of books that provide easy-to-read information about food production and feature fun illustrations of farm life.

www.agclassroom.org - The Agriculture in the Classroom National Resource Directory is an online searchable database that lists hundreds of educational resources designed to help educators locate high quality classroom materials and information to increase agricultural literacy among their Pre-K through 12th grade students.

www.kellyhahnphotography.com - Kelly Hahn Johnson is an award-winning photojournalist with more than 20 years of experience. See more photos, videos and testimonials here.

www.rebeccalongchaney.com – View photos, testimonials and video footage of Rebecca, Rianna and Sheridan's mission to teach young people about agriculture. Order more books here.

The Chaneys raise Hereford beef cattle. Rebecca, Rianna and Sheridan have been honored to take their agricultural message on the road. The trio has visited schools and fairs giving 45-minute agricultural presentations helping thousands of children learn about production agriculture.

Lee and Rebecca Chaney are proud their twin daughters continue a family tradition and passion for agriculture.

Being the fifth generation farmers, the girls enjoy raising their 4-H project steers, pigs, chickens and rabbits. They also love Sunday School, playing softball, riding their bike and checking cows with their Daddy. For more information about their ag presentations, or to order books, contact her at chaneyswalkabout@aol.com, info@rebeccalongchaney.com or check out her website at www.Rebeccalongchaney.com.

Libbi, Lindsi and Landri are proud seventh generation ranchers and enjoy raising animals with their parents Trent and Kelli Loos. Libbi and Lindsi show cattle and pigs in 4-H. They also enjoy many other activities such as softball and volleyball.

The girls help their parents on a ranch in central Nebraska where they specialize in purebred Limousin and Angus cattle, meat goats and purebred hogs. The entire family shares a passion for agriculture and for spreading the word about the many positive things in production agriculture. Kelli is active in several cattle women and agricultural groups and Trent is a nationally known agricultural advocate traveling the country, meeting with farmers and ranchers, and then sharing their stories on his daily radio programs "Loos Tales" and "Rural Route," heard in 19 states and worldwide on the internet.

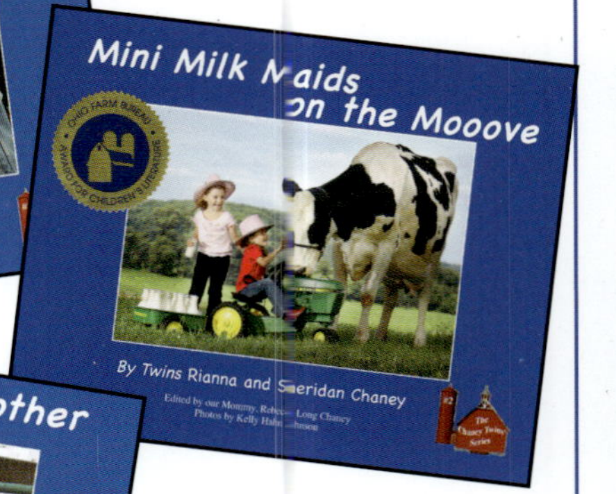

The Chaneys are proud their first three children's books have received numerous "Agricultural Book of the Year" recognitions!! The "Chaney Twins' Series" and lesson plans have been added to the USDA Agriculture in the Classroom (AIC) Resource Directory! They were featured resources on the USDA AIC website for May, 2012.

Lesson Plans are now available free in downloadable form on the Pennsylvania Beef Council website at www.pabeef.org

Box book discounts are available!

Kathy Moser Stowers lives with her husband, Tom and son, Wade on her family's second farm, Twin View Acres 2, near Jefferson, Maryland.

She has 25 years of experience in design and layout at The Frederick News-Post. She has shared her talents and expertise with local, state, national and international groups, organizations and events.

For aide in designing brochures, pamphlets, program books, business cards, etc., contact Kathy at wakstowers@aol.com or call (301) 748-9112 or (301) 371-9306.

Award-winning photographer Kelly Hahn Johnson not only is known for her unique photojournalism style but also her approach to portraiture. She's won countless awards and her images have been featured in local, state and national publications. She loves observing people, moments and emotions, creating photos that will be treasured for years to come, like the priceless images she's captured of the Chaney twins over the years. She lives with her husband Blane and son Brady in a century-old renovated house in Sharpsburg, Maryland.

Visit her online gallery at kellyhahnphotography.com. Contact Kelly at info@kellyhahnphotography.com or call her at (240) 285-3677.

A special thank you to

Trent, Kelli, Libbi, Lindsi, and Landri Loos for a wonderful week and experience on their ranch.